50 WORDS

ABOUT NATURE

BIRDS

TARA
PEGLEY-
STANGER

DEBBIE
POWELL

OXFORD
UNIVERSITY PRESS

Note to Grown-ups

Learning lots of new words is a wonderful aid for young children's language development. A wide vocabulary also helps children to explore and understand the world around them as they grow and learn. Reading the words while looking at the pictures together creates a valuable learning experience.

This book includes new words as well as familiar ones. Even grown-ups might not know some of the words, and there is a pronunciation guide at the end of the book to help.

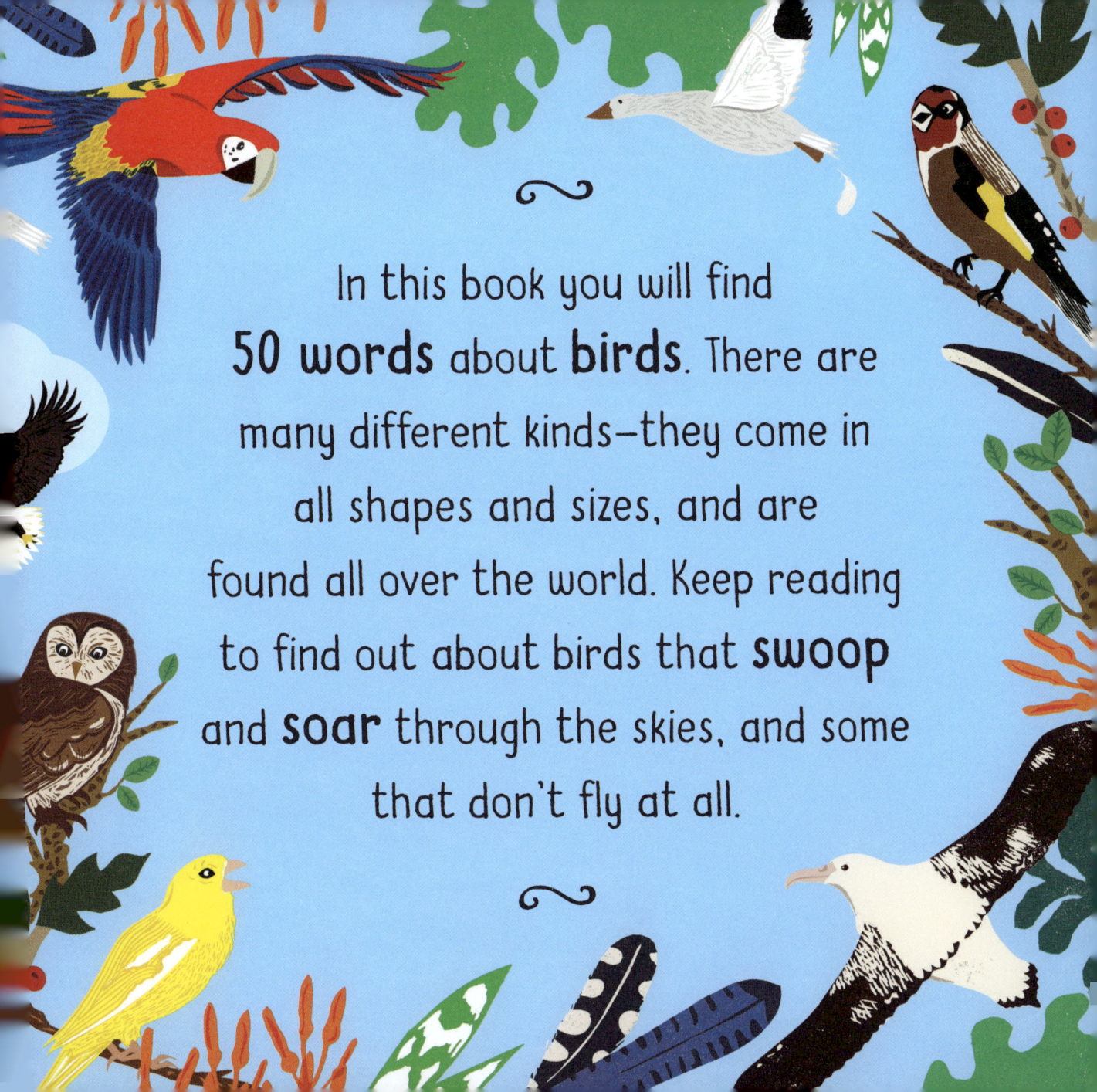

In this book you will find
50 words about **birds**. There are
many different kinds—they come in
all shapes and sizes, and are
found all over the world. Keep reading
to find out about birds that **swoop**
and **soar** through the skies, and some
that don't fly at all.

Birds are **vertebrates**, or animals with backbones. They have:

a **beak**, sometimes called a **bill** ...

... **feathers** ...

... **wings.**

Most birds use
their wings to fly.

This colourful bird is
a **scarlet macaw.**

All birds hatch from **eggs**.

Many birds build a **nest** for their eggs to keep them safe.

A group of eggs is called a **clutch**.

Adult birds lay eggs, then sit on them to keep them warm. This is called **incubation**.

Canary egg

Raven egg

Birds' eggs are not all the same size, shape or colour...

Ostrich egg

(the biggest egg of all, more than 20 times the size of a chicken egg!)

Chicken egg

Emu egg

Waterfowl are birds that live near ponds, rivers and lakes.

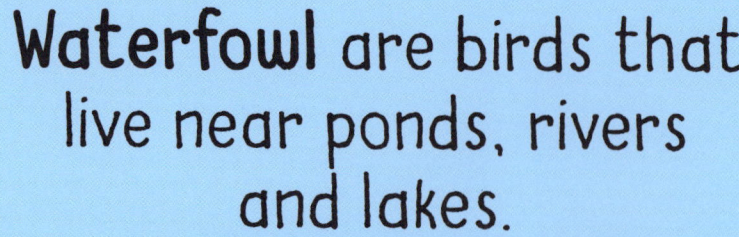

They are good swimmers, and have **webbed feet** to help them paddle along.

They also have **waterproof** feathers.

Ducks and **swans** are waterfowl.

Ducklings are baby ducks ...

...and **cygnets** are baby swans.

Seabirds live near the ocean, where they feed on fish and other sea creatures.

This one is a **wandering albatross.**

This bird spends
most of its time swooping
across the ocean on its
huge wings.

It has the widest
wingspan of any bird.

The **bee hummingbird** is the smallest bird in the world.

It is small enough to fit in your hand.

Hummingbirds have a long beak for feeding on **nectar**, the sweet liquid inside flowers.

The bird's wings move very quickly, making a humming sound.

Hummingbirds are the only birds that can fly backwards, and even upside down.

Songbirds make a range of musical sounds.

canary

This canary sings especially loud and beautiful songs.

Different types, or **species**, of bird sing different songs.

All songbirds are **perching** birds.

This means that their feet can hold on to a **perch**, such as a branch.

goldfinch

The
Asian peacock
is a male
peafowl.

It has
a **crest** on
its head...

...and very
long tail feathers
that trail behind it.

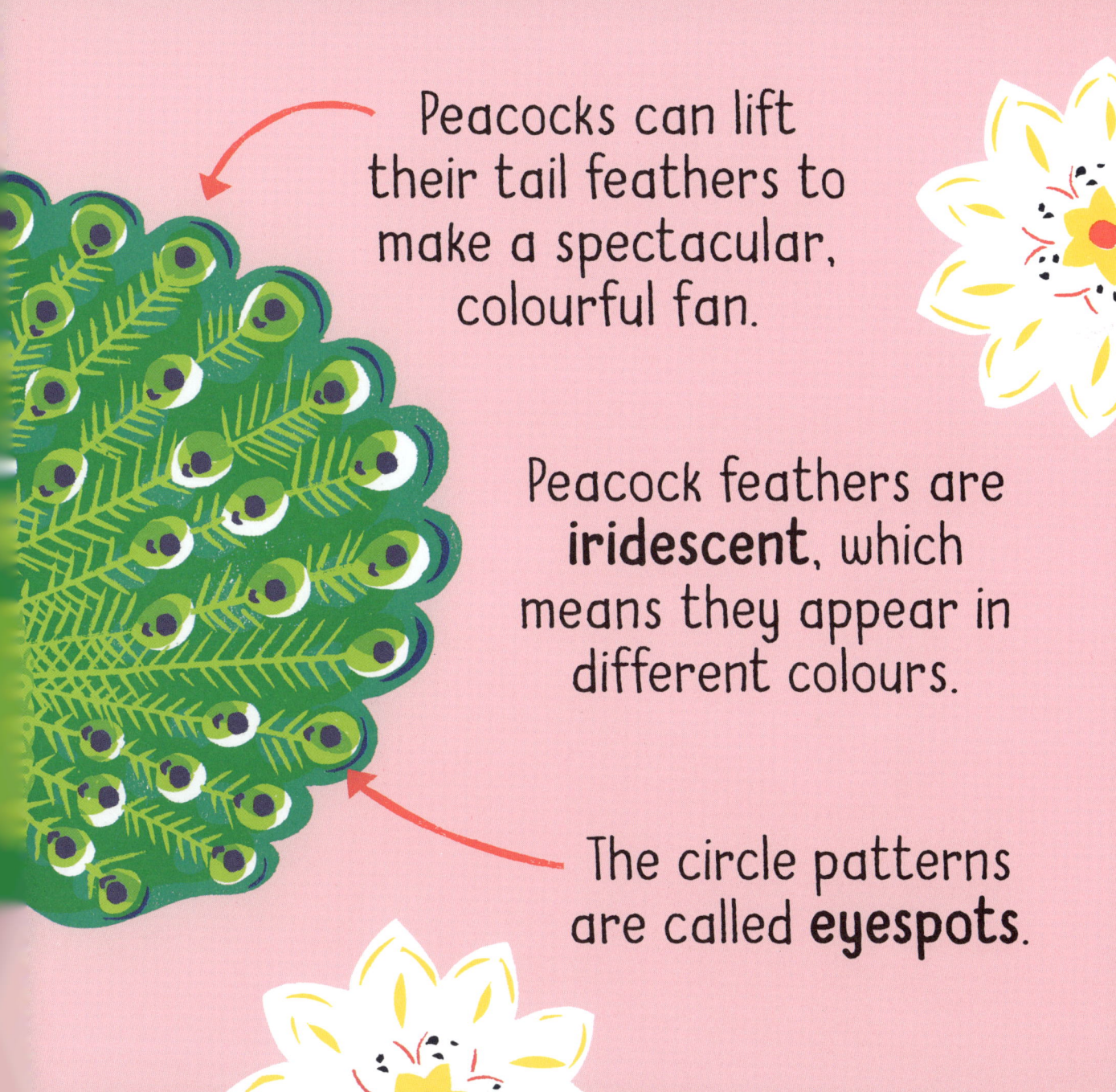

Peacocks can lift their tail feathers to make a spectacular, colourful fan.

Peacock feathers are **iridescent**, which means they appear in different colours.

The circle patterns are called **eyespots**.

Some birds are hunters, known as **birds of prey**, or **raptors**.

This **bald eagle** is one of the biggest raptors.

It has a hooked beak…

...powerful wings
and legs...

...and
sharp claws,
or **talons**.

It hunts animals including fish,
birds, squirrels and lizards.

Most **owls** hunt for mice and other small creatures at **twilight**, just as the light fades from the sky, or at night.

❧

An owl has large eyes to help it see in the dark, and its head can move almost all the way around.

This is a
tawny owl.

Many birds have a
summer home, and
a winter home,
in a warmer place.

They make
a long journey
between the two
homes twice a year.

This is called
migration.

These **snow geese** fly from Canada all the way to California, in the USA, and Mexico.

All birds have wings,
but not all of them can fly.

Penguins
cannot
fly but their
wings help
them to swim.

The wings act like flippers, which help
them swim very fast through the sea.

Ostriches are also **flightless** birds.

They are the tallest birds in the world, and can grow much taller than an adult human.

People who study birds are
called **ornithologists**.

They might
study how
birds live
and behave . . .

. . . birds' **habitats**,
which means
the places where
they live . . .

...or how we can protect birds.

Some are in danger of dying out, like the wandering albatross, or this beautiful **fruit dove**.

flock

toucan

You could be like an ornithologist and learn more about birds.

❦

How many different kinds of bird can you spot in one day?

❦

How many other words about birds do you know?

roost

swoop

chick

woodpecker